Winter Math Walk

Deanna Pecaski McLennan

For Max, Brooks, and Preston

Copyright © by Deanna Pecaski McLennan
First edition 2019

All rights reserved.

No part of this publication may be reproduced in any form, or by any means, electronic or mechanical, including photocopying, recording, or any information browsing, storage or retrieval system, without permission in writing from the author.

www.mrsmclennan.blogspot.ca

Winter is a peaceful time of year.

We go for long walks on the trail.

If we look closely, we find math everywhere!

High in the tree we spot a pinecone.

Its ends are tipped with snow.

Why are the scales spiralled so tightly?

We discover frost covered branches.

They appear to glow in the sunlight.

How did such interesting crystals form?

Up ahead we walk along the creek.

It has not yet frozen.

What is that mist rising above the water?

We notice a different kind of frost.

This plant appears to be painted white.

Why is there more on its perimeter than area?

We see tracks on the trail.

There is distance between each print.

Who was walking here before us?

I hold out my mitten.

A single snowflake lands in the center.

How is it formed so perfectly?

There is snow settled along the fence.

It is in the shape of triangles.

How many patterns do you see?

We walk towards an old tree.

It is much taller than we are.

Why is the bark different in the middle?

We return to our neighbourhood.

Even the stop sign is covered in snow.

How did the icicles form at the bottom?

We see etchings on our house windows.

The frost looks like tiny feathers.

How did the delicate design appear?

The snowman in our yard greets us.

His happy face makes me smile.

How can we make sure he does not tip over?

Next to the snowman stands our maple.

Its knobby twigs dance in the wind.

Why does each branch look like a mini tree?

We step onto a frozen puddle.

There are bubbles stuck within.

How can we measure the depth of the ice?

After our walk we spend time in our yard.

The snow is so much fun to play with!

How big a fort can we build?

Author's Note

Spending time immersed in nature is a wonderful way for young children to learn about our world.

There is beauty in mathematics. Helping children to recognize the strong connection between nature and math may encourage them to see mathematics as an aesthetic and captivating subject. So often the only experiences children have with math are those found inside the classroom. Exploring the authentic math that exists in nature may help nurture children's interest and confidence, building a strong foundation for subsequent experiences.

My hope in writing this book is to inspire children, educators and families to see math as an inviting discipline that lives beyond the walls of the classroom. Our natural world is filled with amazing mathematical connections. This book does not need to be read beginning to end. The photos can be used individually, or in combination, to spark mathematical conversations and connections with children. Ask children what they see, think and wonder about each picture. Ask what their theories are for what they see happening on each page. At the end of the book you will find information to compliment each photo. Adults can support and extend children's mathematical and scientific ideas using this information. Additional resources can scaffold and build inquiries that spark from the text.

The information presented in this book can serve as an introduction to new math concepts, or as a reference when natural treasures are discovered by children outdoors. Consider reading it together with children before venturing out into the world on your own math walk. You might choose to use the photos as conversation starters, or read the book in its entirety using photos and narrative.

When we look at the world through a mathematical lens, we discover that anything is possible!

~Deanna

In this photo draw children's attention to the tightly wrapped seed scales that grow around the pinecone. Ask children to hypothesize why the scales develop in this way, and if they can notice and name a pattern.

Pine cones are seed pods for pine trees. The cones have thick, woody scales covering them that protect the seeds inside. When it is cold the scales of the pinecone are closed, protecting them from the elements. As the weather warms the scales open and the seeds are released. If you look closely at pinecones, the scales appear to grow in spirals that follow the Fibonacci sequence, a specific numerical pattern. This helps the plant optimize seed growth.

Math ideas might include size, shape, patterns, temperature, and counting.

In this photo draw children's attention to the delicate crystals that have formed on the branches. Ask children to hypothesize why the frost has developed in this way, and if they have mathematical observations about its formation along the lengths of the branches.

Hoar frost looks like tiny ice crystals and forms when outdoor objects cool below the frost point of the surrounding air on cold, clear nights. Hoar frost can also form in manmade objects including freezers if they are not insulated properly.

Math ideas might include temperature, shape, measurement, length, and patterning.

In this photo draw children's attention to the mist rising over the water in the background of the photo. Ask children to hypothesize what the mist might be, and why it would be over the water on a such a cold day.

Water vapour is invisible. It is composed of tiny water droplets that scatter light, giving it a misty appearance. When warm water evaporates directly into a cold air mass above it, the mist forms. Evaporation fog can frequently be observed in late fall and early winter. This can happen even when the water and air are very cold, including winter time.

Math ideas might include temperature, properties of materials, size, and shape.

In this photo draw children's attention to the frost that has formed on the individual leaves. Ask children to hypothesize why the frost on the perimeter of each leaf appears to be more noticeable and prominent than the frost inside the area of the same leaf. Ask children to compare and contrast this type of frost to the hoar frost that formed on the branches earlier in the book.

Advection frost forms when cold wind blows over outdoor surfaces. It appears as small spikes on objects. Frost forms at night when it is colder, and appears to disappear or evaporate once the sun rises and the temperature warms.

Math ideas might include area, perimeter, temperature, classification and properties of materials.

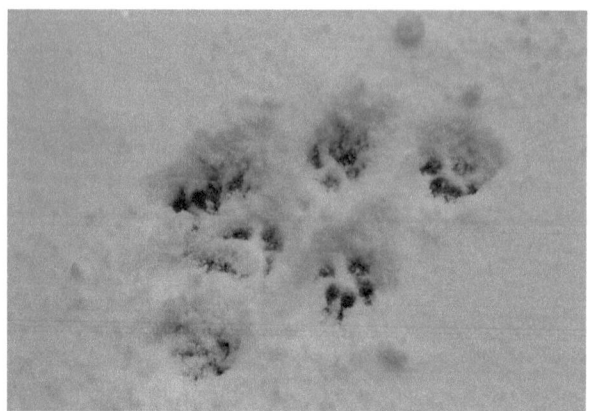

In this photo draw children's attention to the footprints that are visible in the snow. Ask children to share their experiences finding, researching, and identifying the prints they have previously discovered.

Footprints can give clues about what type of animal has travelled through. Details such as the size and shape of the print, the number of visible toes, and the length of the animal's claws all help with identification. Footprints are frieze patterns. Frieze patterns are designs on a flat surface that move in one direction and are repetitive in nature. Measuring the distance between each track gives clues to the animal's gait (e.g., fast/slow, left side/right side).

Math ideas might include shape, size, speed, opposites, measurement, reflection and symmetry.

In this photo draw children's attention to the intricate detail of the individual snowflake located in the center of the mitten. Point out its six sides and rotational symmetry. Ask children to describe what they see, and share connections to their own experiences collecting fresh snowflakes falling from the sky.

Snowflakes begin as water droplets. They freeze into crystals depending on the temperature and atmospheric conditions. The shape of the crystal formed depends on the specific temperature outdoors and includes needles, hollow columns, sector plates, and six-armed dendrites. Most snowflakes are between 1 and 5 cm in diameter.

Math ideas might include size, shape, symmetry, temperature, and patterning.

In this photo draw children's attention to the interesting way the snow has settled in each section of the fence. Ask children to describe what they observe and hypothesize as to why the snow only covered this part of the fence.

Snow is formed when temperatures are low and there is moisture in the atmosphere. The closer it is to the freezing point, the larger the snowflakes will be. If snowflakes fall through air that is slightly above freezing temperature, the edges of the flakes will melt and the flakes will stick to one another, making them appear larger.

Math ideas might include size, shape, patterning, temperature, properties of materials, and symmetry.

In this photo draw children's attention to the interesting textures on the tree's bark. Most is rough and weathered while the damaged part appears to be smooth. Ask children to imagine what might have happened to the tree, and predict how it will fare over the course of the winter. Ask them to consider why the healthier part of the bark is coarse and weathered looking.

A tree's bark is a thick outer covering that protects a tree. Trees can withstand damage of their bark if it is approximately one-fourth or less their circumference. In some cases the tree can be repaired by grafting the damaged area with healthier sections of the tree.

Math ideas might include texture, patterning, time, area, and measurement.

In this photo draw children's attention to the icicles formed at the bottom of the sign. The shape and colour of the sign, as well as the snow covering the top, might also be of interest to them. Ask children to describe what they see, and hypothesize as to how the icicles were formed.

Icicles form when the outdoor air is below freezing, and sunshine warms an object upon which there is snow. In this photo the snow on the sign was warmed by the sun and melted, and as the water dripped down it was cooled by the surrounding air and refroze. The repeated pattern of melting and freezing formed the icicle. With each cycle the icicle grows larger in diameter and longer.

Math ideas might include length, width, temperature and patterning.

In this photo draw children's attention to the intricate frost formed on the window. This is the third picture of frost presented in the book. Ask children to describe what they see, and compare the photo of this frost to the other two photos.

In the winter water vapour inside a home condenses and freezes onto cold window glass in beautiful fractal patterns. A fractal is never-ending and self-similar. This means that a small section looks like a mini version of the larger pattern. Window frost repeats over and over in a loop.

Math ideas might include fractals, patterning, shape, line, and temperature.

In this photo draw children's attention to the number and shape of balls used to make the snowman. Ask children to share their previous experiences making snowmen and what advice they would offer new builders.

Snow that can easily be formed is best for rolling balls used for snow creations. Snow can be classified as wet or dry. If snow is too dry, it is powdery. If it is too wet, it is slushy. 'Packing snow' is snow that is near the melting point and the perfect consistency for building. Surrounding air temperature and crystal structure determine the moisture level in snow.

Math ideas might include shape, size, temperature, symmetry, and measurement.

In this photo draw children's attention to the structure of the tree. Point out the many branches visible in the photo. Ask children to share their mathematical observations and describe any patterns they see.

Much like window frost, many trees are also self-similar. Self-similar objects contain patterns. The whole has the same shape as many of its smaller components. A small section of this tree looks like the entire tree, just in miniature. Trees are fractals. These patterns repeat smaller and smaller copies of themselves over and over.

Math ideas might include fractals, patterns, self-similarity, and measurement.

In this photo draw children's attention to the characteristics of the ice - the colours, textures, and markings within it. Ask children to share their experiences with ice, and what strategies they would use to discover how thick it is. Ask them to hypothesize why there are bubbles frozen within the ice?

Most materials, including water, expand when they are heated and shrink when they are cooled. However water shrinks only until the temperature of 4 degrees, and then it expands again, which is why the top of a puddle freezes first. This layer of ice acts as insulation and slows the process of the water freezing below.

Math ideas might include temperature, colour, pattern, texture and measurement.

Deanna Pecaski McLennan, Ph.D., is an elementary educator in Ontario, Canada. Deanna is fascinated by math and loves exploring its natural and authentic application in the living world. She hopes to help children and families recognize math as a beautiful and fascinating subject, and grow children's confidence, accuracy and interest in math.

Follow Deanna on Twitter and Instagram for regular updates including ideas for engaging children in playful, emergent math inside the classroom and beyond. Extending math learning outdoors is a favourite exploration!

Connect with Deanna:

deannapecaskimclennan@gmail.com
@McLennan1977
www.mrsmclennan.blogspot.ca

Also from Deanna

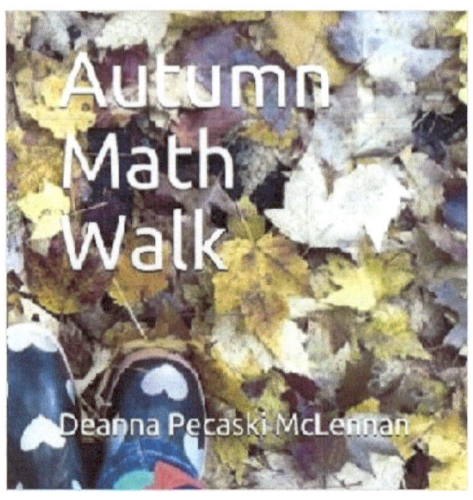

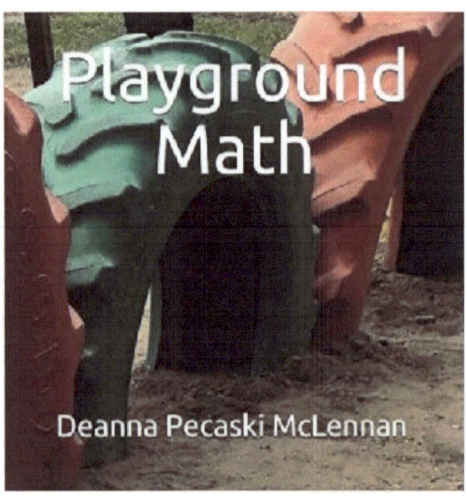

Joyful
Math

www.ingramcontent.com/pod-product-compliance
Lightning Source LLC
Chambersburg PA
CBHW051222220526
45473CB00003B/1137